IRWIN MATH

MORSE, MARCONI AND YOU

D0596549

CHARLES SCRIBNER'S SONS   NEW YORK

I would like to express my thanks to Mr. Hal Keith for the fine work done on the illustrations for this book. I would also like to thank my son Robert and daughter Nicole for their help in trying out all of the projects.

Additional thanks to Mr. Jerry Holtzman of the Bronx High School of Science for his comments and suggestions and to Jack Magaril for his photographs.

I.M.

Copyright © 1979 Irwin Math

Library of Congress Cataloging in Publication Data
Math, Irwin.
  Morse, Marconi and you.
  Bibliography: p. 78.
  Includes index.
    1. Telegraph—Apparatus and supplies—Design and construction.   2. Telephone—Apparatus and supplies—Design and construction.   3. Radio—Apparatus and suplies—Design and construction.   I. Keith, Hal.
II. Title.
TK5501.M37   621.38   79-1351
ISBN 0-684-16081-1

This book published simultaneously in the
United States of America and in Canada—
Copyright under the Berne Convention
All rights reserved. No part of this book
may be reproduced in any form without the
permission of Charles Scribner's Sons

1 3 5 7 9 11 13 15 17 19 M/C 20 18 16 14 12 10 8 6 4 2

Printed in the United States of America

To Ellen,

My wife and best friend

CONTENTS

list of illustrations

foreword

When Morse, Marconi, Alexander Graham Bell, and other inventors were developing the communications devices that revolutionized our world, they used equipment and apparatus built from ordinary materials close at hand. Though the early inventions were crude, they worked—and well enough to demonstrate clearly the principles behind them.

During the late 1930s and '40s many young people were captivated by the emerging field of communications. Their excitement was heightened by the publication of many "learn-by-building" books which encouraged youngsters to build a variety of devices, from simple telegraph equipment to telephone systems and even shortwave radios. There were no ready sources of components for these young experimenters, however; common household items such as wood, tin cans, and bell wire were the materials used, and the technology was "cut and try."

Today many companies offer complete kits of parts with which you can build everything from a burglar alarm to a color television set. And the devices work—but why? The "why" is often passed off by the kit manufacturer in a hastily written paragraph in an instruction booklet; without a true understanding of what has been built, the experimenter's interest inevitably dwindles.

Morse, Marconi and You is meant to bridge the gap between two methods of building, to take advantage of the availability of certain simple components without overlooking the need to understand them. The successful young experimenter of today has as much need to understand the function of every component he uses as did Morse and Marconi.

The reader will learn to build telephone, telegraph, and radio sets, using much the same materials and techniques employed by the original inventors. Most of the components are built of wood, tin-can metal, bell wire, and common hardware store items. Where commercial components are used, their function is clearly explained. When a project is finished, the builder understands all aspects of the device and knows what makes it tick. All the devices described work, and one project leads to the next in logical sequence, often making use of components built for an earlier project.

It is the author's hope that *Morse, Marconi and You* will serve to spark interest in what can become one of the most exciting and satisfying hobbies a young person can have in this electronic age.

I.M.

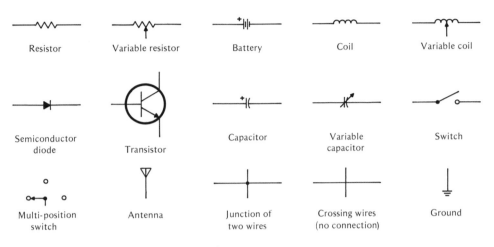

Resistor	Variable resistor	Battery	Coil	Variable coil
Semiconductor diode	Transistor	Capacitor	Variable capacitor	Switch
Multi-position switch	Antenna	Junction of two wires	Crossing wires (no connection)	Ground

Electronic symbols used in this book

CHAPTER 1

EARLy COMMUNICATIONS

Today, every corner of the world is within communicating range. Signals speed through wires across nations, under the oceans, and via satellites far above the earth, relaying thousands of messages to all parts of the globe instantaneously. In laboratories, scientists work with lasers and optical fibers to further increase the number of messages that can be sent. The average person sees events on TV that are taking place on the other side of the world, live and in full color!

It was not always so. Ever since humans first learned to talk there has been a need to communicate over distances that could not be covered with a shout. Even primitive peoples developed methods that worked well for them and, in some cases, work for their modern descendents.

One popular method, practiced in many parts of the world for thousands of years, was to strike the ground with a special hammer. The sound waves produced by the blows traveled through the ground in all directions from the "transmitter." To receive these signals one had to place one's ear to the ground, within range, and listen carefully. Of

FIGURE 1

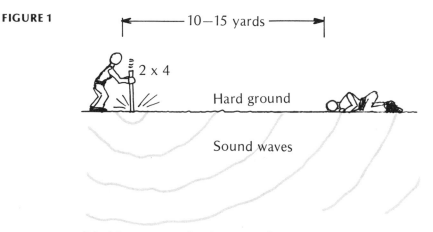

Primitive communications experiment.

course one had to know what the thumps and bumps meant, but this assured privacy.

This ancient system operated over a surprising distance; it is worth doing a simple experiment to see how it works (figure 1). All that is necessary is a 1–2-foot length of 2 x 4-foot lumber and a friend.

In an open field or backyard, preferably with a hard dirt surface, have a friend place his ear directly on the ground, 10–15 yards away from you. Strike the ground quite hard with the 2 x 4. Your friend will hear each blow clearly. Try to increase the distance between "transmitter" and "receiver" and see how far you can send a signal. It will help if your friend tightly covers his free ear to cut out the aboveground sounds.

Africans employed a similar method using a convenient river or stream instead of the ground. Hollow logs were set up in the water. When a log was struck with a hammer, sound waves traveled through the water to another log where they could be heard by a listener. Since the stations were identical, they could transmit and receive; elaborate communication networks were formed, some of them in use well into the 20th century.

This "water telegraph" can be duplicated (figure 2) if you have a stream or lake available—or even a bathtub! Obtain two 1-foot lengths of pipe, such as common drainpipe, and immerse both pieces one-third of

FIGURE 2

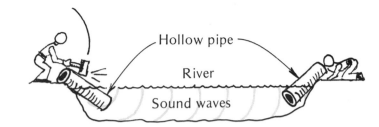

African "water telegraph" experiment.

the way into the water a short distance apart. Have a friend place his ear tightly over the open end of one pipe while you strike the other pipe with a pebble. The sound will be clearly audible to your friend. Again, covering the free ear of the receiver will help minimize stray sounds. Africans have sometimes been able to use this system to communicate over several miles. If a large body of water is available, you will be surprised at how far the sound will travel.

American Indians developed a long-range communications method that is familiar to almost everyone: the smoke signal. By controlling the release of large billows of smoke from a smoldering fire, coded messages could be sent as far as the eye could see. A similar long-distance system was operated at night by alternately blocking and unblocking the blaze of a fire, usually on top of a high hill.

These and other schemes using oil lamps, drums, flags, semaphores, and the like were the main forms of long-distance communications until the perfection of the electric telegraph by Samuel F. B. Morse in 1832. With this invention signals could be sent wherever wires could be strung—even under the Atlantic Ocean. Morse's telegraph revolutionized communications. Several years later the invention of the telephone by Alexander Graham Bell allowed speech to be sent over telegraph wires, and the era of modern communications began. In 1895 Guglielmo Marconi succeeded in transmitting electric signals between two places without the use of wires, thereby removing the last obstacle to worldwide communications.

CHAPTER 2

LET'S GET down TO bASICS

No one has ever seen electricity, but learning to use it properly need not be difficult. Understanding a few basic principles can make the difference between the successful experimenter and the disappointed amateur. Let us begin with a few facts that will be useful in our future experimenting.

Always remember that electricity flows — just like a liquid. The flow of electricity through a wire can be compared to the flow of water through a pipe. If you drilled a hole in the side of a pipe (figure 3) and looked in, you would see that a certain amount of water was passing the viewing point each second, producing a water current like the current in a

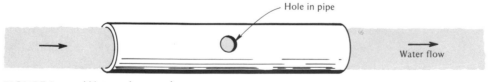

Hole in pipe

Water flow

FIGURE 3   *Water pipe analogy.*

4

river. Stick a finger in the hole in the pipe and you will feel the pull, or strength, of the current. This strength is determined by how much force or pressure is being applied at the end of the pipe.

A wire with electricity flowing through it has similar characteristics. The flow of electricity is called electric current. The strength of this current is also determined by the amount of pressure applied to the wire—in this case, electric pressure.

In the water system, the current is measured in gallons moving through a cross section per second. Pressure is measured in pounds per square inch. Electric current is measured in units called amperes and pressure is measured in volts. In either system, the higher the pressure, the greater the current.

It is not enough for an electric current to flow to have pressure (voltage) available. The current must flow in a complete path, or circuit, from its source through wires and other components, and back to the source. A break in the circuit will stop the flow.

Figure 4 shows a simple electric circuit where the current flows from the battery through the switch (when it is closed), through the light bulb, and back to the battery. If you break the circuit by opening the switch, or if there is a break in the wires, the flow of current will stop and the bulb will go out. (When a light bulb burns out, the circuit inside the bulb is interrupted.)

FIGURE 4

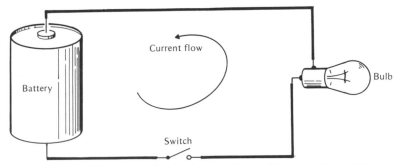

Flashlight circuit. This is the basic circuit used in most flashlights and lanterns.

FIGURE 5

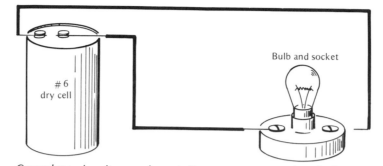

Complete circuit experiment. Be sure to remove insulating covering from wire.

Whether the circuit is a simple flashlight or a complicated hookup from an electric power station to your home, the circuit must be complete—which is why the power cord connected to a home appliance has at least two wires, one for incoming flow, the other for outgoing flow.

A simple experiment will demonstrate this. A few common items, which are easy to obtain, are needed: a #6 dry cell, a piece of common bell wire, a #222 flashlight bulb, and a matching socket from your hardware store. Carefully remove some covering from each end of two 12-inch lengths of wire and hook up the circuit shown in figure 5, and connect one wire to the battery. The bulb will not light, because there is no complete path. Now connect the open wire to the remaining screw on the socket, and the bulb will light.

Electric current will only flow through certain materials, called conductors. Almost all metals, particularly copper (the metal most wires are made of), are good conductors. A material that electricity cannot flow through is called an insulator. The covering on the wire used to hook up the light-bulb circuit is an insulator. It is used to keep the current in the wire and prevent it from flowing to adjacent wires if the wires accidentally touch.

The light-bulb circuit can easily be modified to help us learn about which materials are conductors and which are insulators. Cut another 12-inch length of wire and reconnect the circuit (figure 6). The bulb will not light with the two wires unconnected. If a conductor of electricity is

FIGURE 6

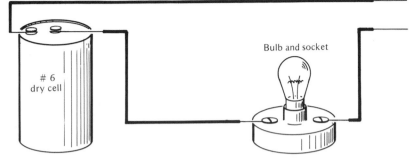

Insulator/conductor tester. With this simple device you can check the conductivity of most materials. You can also use it to be sure that electrical connections have been properly made.

placed between the wires, the circuit will be completed and the bulb will light. Using this simple conductivity tester, we can experiment with glass, plastic, wood, rubber, and various metals to determine which are conductors and which are insulators.

Not all conductors are metals. Liquids such as water can also conduct electricity. To see this, place the two wires from the tester in a glass of water. Water is a conductor, but its degree of conduction is too low to light the bulb. Now add 1–2 tablespoons of salt to the water. The dissolved salt increases conduction to the point where the bulb will light.

The battery in our experiment has a pressure of only 1.5 volts. If the voltage is much higher than a small battery can supply (say, the 115 volts from a household outlet), plain water can not only light a bulb but shock and even seriously injure a person standing in a puddle of water if that person touches a faulty electric appliance.

Electricity can be produced in many ways. The experimenter often uses the common dry-cell battery (shown cut apart in figure 7), consisting of a carbon rod, a zinc can, and moist chemical filler. It is the interaction of these that produces electricity. A carbon-zinc dry cell has a voltage of 1.5 volts regardless of its size, whether it is our #6 cell, the familiar D cell, or a tiny AAA penlight cell. But the large #6 cell will deliver almost 75

FIGURE 7

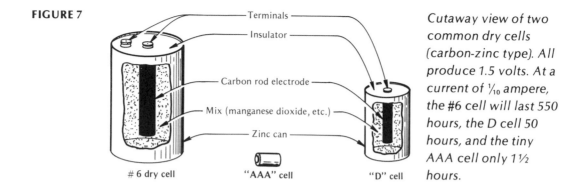

Cutaway view of two common dry cells (carbon-zinc type). All produce 1.5 volts. At a current of ⅒ ampere, the #6 cell will last 550 hours, the D cell 50 hours, and the tiny AAA cell only 1½ hours.

Terminals

Insulator

Carbon rod electrode

Mix (manganese dioxide, etc.)

Zinc can

6 dry cell   "AAA" cell   "D" cell

times as much current as the AAA cell. In a fixed current application, such as keeping a lamp lit, the larger battery will do the job longer.

Many cells can be connected together to produce high-voltage batteries (figure 8).

Wet-cell batteries such as the lead-acid battery used in automobiles (see simplified form in figure 9) are common. The electricity in a lead-acid battery is produced by the action of an acid-water mixture on two plates, each made of a different metal. Other chemical batteries may use different metals, alkaline mixtures, and so on, but the common fault remains: after a while the chemicals are used up, to the point that voltage and cur-

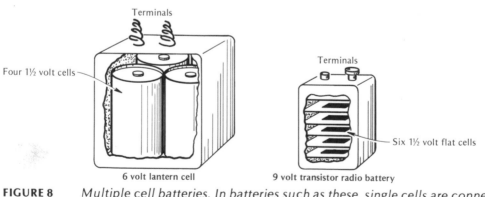

Terminals

Four 1½ volt cells

Terminals

Six 1½ volt flat cells

6 volt lantern cell   9 volt transistor radio battery

FIGURE 8   *Multiple cell batteries. In batteries such as these, single cells are connected together to form higher-voltage units.*

FIGURE 9

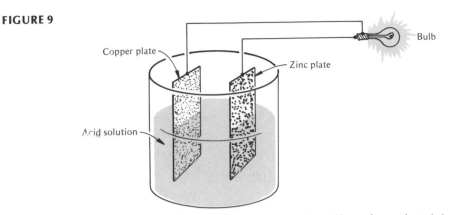

Simple "wet" cell. Most dissimilar metals will produce electricity

rent decrease and make the battery useless. This is why batteries are only useful for relatively short-lived applications where current from a stationary source is unavailable.

Where continuous power is needed, as in homes, schools, and factories, the method used to produce electricity is based on the fact that magnetic energy and electric energy are easily converted to each other.

To illustrate this we need a bar magnet of the kind supplied with some children's games and a Boy Scout-style compass. The delicately balanced compass pointer will be our electric indicator, since a light bulb requires much more current than this experiment produces.

To convert the compass into an indicator, wind a coil of five turns of wire (figure 10) and hold it in place with masking, adhesive, or other tape. Slip the coil over the compass, making sure you can still see the pointer. Wind another five-turn coil of wire as part of the electric "generator" and connect it to the compass coil. The two coils are connected in a complete circuit.

To generate electricity, move the bar magnet quickly in and out of the "generator coil." The compass needle will deflect each time you move the magnet, proving that an electric current is flowing. The moving magnet is converting magnetic energy into electric energy, which flows through the connecting wires to the second coil, where it is converted

FIGURE 10

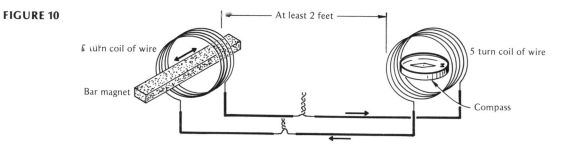

Generator experiment. Moving the bar magnet in and out of the coil will cause a current to flow in the circuit, deflecting the compass needle.

back into magnetic energy and deflects the compass needle. We will look at these conversions in more detail in chapter 3.

By means of other sources of power, such as water or steam, huge magnets can be moved in close proximity to large coils of wire, generating large amounts of electricity. Such commercial generators operate as long as external mechanical power is applied to them. This is how we obtain the continuous energy required by our society.

CHAPTER 3

THE TELEGRAPH

We already know that by moving a magnet near a coil of wire we can produce a flow of electric current in the wire. By the same principle, when an electric current flows in a wire it produces a magnetic field around the wire: this is why the compass needle was deflected in our "generator" experiment.

The magnetic field produced around a single wire is usually quite small. If the wire is coiled up, the small amount of magnetism from each turn reinforces the magnetism from the next turn, resulting in a stronger magnetic field. Now, if the wire is wound on an iron core, the magnetic field will be concentrated in the core, instead of around the wires; it will be present at the ends of the core where we can make use of it.

To demonstrate this, we will again require a #6 dry cell, some wire, and a large iron nail. Wind twenty turns of bell wire around the nail (figure 11). Connect the free ends of the coil to the battery. The nail will become strongly magnetized and will pick up other nails and pieces of iron.

FIGURE 11

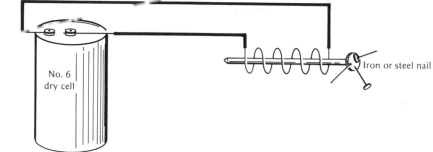

Electromagnet experiment. The strength of the magnet depends on the number of turns of wire and the number of batteries used.

When the coil is disconnected, stopping the current flow, the magnetic field will disappear and the nails will drop. Do not keep the circuit energized for too long: the circuit requires a large amount of current from the battery and will quickly wear the battery out.

This simple experiment illustrates the basic principle that led Samuel F. B. Morse to the invention of the electric telegraph. For over forty years, until the invention of the telephone, the telegraph was the most important long-distance communications instrument.

Figure 12 shows how the experiment just performed can lead to a communications device. We arranged a strip of spring steel close to a magnetic coil or electromagnet. When the coil is energized, the strip is pulled to the electromagnet. As it strikes the iron core, a click is heard.

FIGURE 12

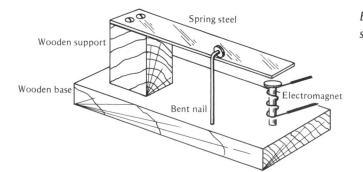

Basic telegraph sounder.

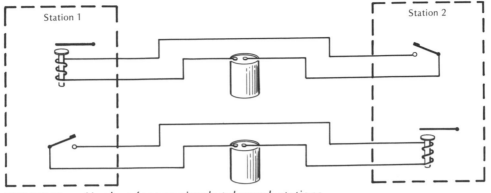

FIGURE 13   *Hookup for two simple telegraph stations.*

When the current flow is interrupted, the steel returns to its initial position, striking the bent nail with another click. If the current is interrupted twice, quickly, the two "dots" can be assigned to represent a letter of the alphabet (the letter *i*, in the international Morse code). Thus a series of clicks can represent a word. By placing the current-controlling switch at a distance from the electromagnet, we can send messages over that distance, letter by letter—which is, in fact, what the telegraph does.

Figure 13 shows an electric wiring diagram of two circuits arranged to form a two-way telegraph system. The electromagnet spring units are called sounders, and the current-interrupting switches are called keys. A code of clicks representing the letters of the alphabet, the numerals from 0 to 9, and the most common punctuation marks was established by Morse and is known as the international Morse code (tabulated in figure 14). Dots signify a short interval between the two clicks; dashes, a longer interval. The letter *a* (dot-dash) would sound like "click click (long interval) click (pause) click." The international Morse code is also used in shortwave radio communications, with dots represented by short-interval tones and dashes by longer-interval tones. For example, the letter *a* would be bz (pause) bzzzzz. Readers planning significant experimenting in communications should learn the code by heart. Studying along with a friend will make mastering the code easier.

FIGURE 14

A · —	N — ·	1 · — — — —
B — · · ·	O — — —	2 · · — — —
C — · — ·	P · — — ·	3 · · · — —
D — · ·	Q — — · —	4 · · · · —
E ·	R · — ·	5 · · · · ·
F · · — ·	S · · ·	6 — · · · ·
G — — ·	T —	7 — — · · ·
H · · · ·	U · · —	8 — — — · ·
I · ·	V · · · —	9 — — — — ·
J · — — —	W · · —	10 — — — — —
K — · —	X — · · —	Period · — · — · —
L · — · ·	Y — · — —	Comma — — · · — —
M — —	Z — — · ·	Question mark · · — — · ·

The international Morse code. A dot signifies a short tone or two rapid clicks in telegraphy. A dash signifies a long tone or two slow clicks.

You can communicate with a friend (and speed up your learning of international Morse code) if you have a complete telegraph set. A simple but practical and reliable key and sounder can be built by anyone with moderate skills (figure 15). The sounder is built like the one just discussed and is mounted on the same base as the key. There is also a send/receive switch which allows two similar units to be hooked up for two-way communications. Step-by-step instructions for building the sets follow.

FIGURE 15

A complete telegraph system that can be easily built by anyone with a little mechanical ability.

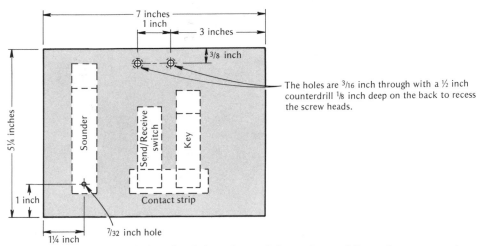

The holes are 3/16 inch through with a ½ inch counterdrill ⅛ inch deep on the back to recess the screw heads.

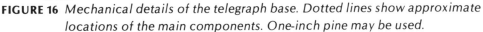

FIGURE 16 *Mechanical details of the telegraph base. Dotted lines show approximate locations of the main components. One-inch pine may be used.*

BASE CONSTRUCTION (figure 16)

1. Cut a piece of common 1 x 6-foot pine, 7 inches long. Sand all surfaces smooth, particularly the cut edges. If desired, a bevel may be cut all around for better appearance.

2. Carefully mark the hole pattern on the pine (figure 16). Measure all dimensions twice to be sure they are correct.

3. Drill all holes with a hand power drill. Counterdrill all holes from the rear.

4. Give the completed base a coat of varnish or shellac to seal and protect it.

ELECTROMAGNET CONSTRUCTION (figure 17)

1. Obtain a ¼ x 3-inch stove bolt, two ¼-inch washers, and a ¼-20 nut from a hardware store. Assemble the nut and washers as shown in (A).

FIGURE 17

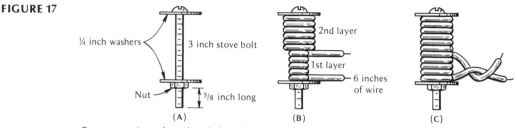

Construction details of the electromagnet.

2. Wind a single layer of black plastic electric tape between the two washers.

3. Wind four layers of #26 plastic-covered wire (PVC) between the two washers in neat, even layers. Start each layer directly over the last turn of the layer just completed: thus, alternate layers go in opposite directions. Be sure the wire is tightly wound. Begin the first layer from the nut end; (B) shows what the electromagnet looks like after 1½ layers have been wound.

4. After the fourth layer is wound, twist the two wires together to hold the layers in place. Give the coil a coat of varnish and let it dry thoroughly.

5. Mount the electromagnet in the $\frac{7}{32}$ hole in the base by screwing it in with a screwdriver. Tighten the screw, which will cut its own threads, until the assembly is secure; but do not overtighten.

SOUNDER CONSTRUCTION (figure 18)

1. Very carefully cut a strip of metal from a used tin can. Use tin snips or a heavy-duty pair of scissors. If the can is flattened first, the task will be easier. Get the dimensions as close to those shown in the figure as possible. *Sand all edges, as the metal is very sharp and you can easily get cut if you are careless.* It is a good idea when cutting tin-can stock to wear a pair of inexpensive work gloves.

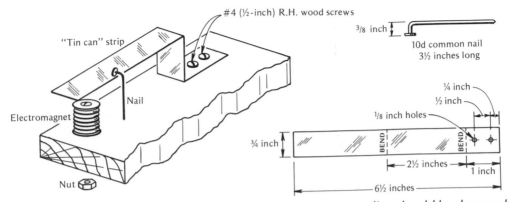

FIGURE 18 *Sounder construction and location on base. Bending should be done only after the ⅛-inch holes are drilled and all sharp edges are smooth. Be careful not to get cut. Tin-can stock can be very sharp!*

2. Carefully drill two holes in the metal strip and draw the two bending lines.

3. Bend the metal strip in two places in a vise, or by grasping with two pliers.

4. Mount the strip on the base with #4 (½-inch) round-headed wood screws. Make sure the strip lines up directly over the electromagnet.

5. Bend a tenpenny finishing nail (10d) as shown into the base, then drive so it just contacts the strip without pressing on it.

6. Thread a 1⅛-inch machine screw and nut into one of the terminal holes on the base and connect an electromagnet lead to this screw under the nut. Don't forget to remove the insulation from the wire. Tighten the screw and nut securely; add another nut loosely to be used as a terminal.

KEY CONSTRUCTION (figure 19)

1. From the same tin can used for the sounder, cut a strip of metal to the size indicated on the drawing.

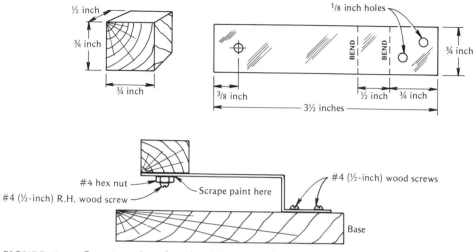

FIGURE 19   *Construction details of key. Bending should be done after holes are drilled.*

2. Carefully drill the three holes shown; draw the bending lines, and bend the strip of metal.

3. Cut a wooden knob from pine, or attach a small kitchen drawer knob with a #4 (½-inch) wood screw and nut. Be sure the screw is very tight, and sand off any paint on the underside of the metal strip.

SEND/RECEIVE AND CONTACT STRIP CONSTRUCTION (figure 20)

1. Cut a strip of metal for the contact strip from the tin can used for the sounder.

2. Carefully drill the two holes indicated, and mount the contact strip with a #6 sheet metal screw directly into the base. Put in the right-hand screw first.

3. Put a wire from the electromagnet under the contact strip, after removing the insulation; then install the other #6 sheet metal screw. Be sure both screws are tight.

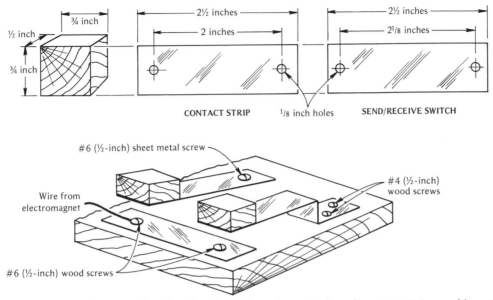

FIGURE 20 *Construction details of send/receive switch and contact strip, and locations on base.*

4. Place the key so that its screw (holding the knob) lines up just over the right-hand screw in the contact strip.

5. Install two #4 (½-inch) round-headed wood screws, but tighten only one.

6. Cut another strip of metal from the tin can as you did in constructing the key.

7. Drill two holes.

8. Make another wooden knob and attach it with a #4 (½-inch) wood screw and nut as in constructing the key. Securely tighten the screw, sandpapering the metal under the screw to assure good contact.

9. Mount the switch metal strip with a #6 (½-inch) sheet metal screw and two washers. It should line up over the left screw of the contact strip.

10. Tightly screw another 1⅛-inch terminal screw into the base and loosely thread a 6–32 nut on it.

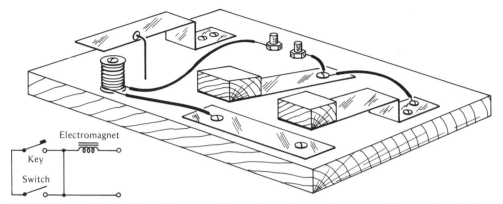

FIGURE 21 *Final wiring diagram of the telegraph set. You may wish to paint the key knob and send/receive switch knob different colors to avoid confusion.*

FINAL WIRING (figure 21)

1. Connect a wire from the metal strip used for the switch to the metal strip used for the key.

2. From the same point, connect another wire to the remaining terminal screw (under the nut). Tighten the nut securely.

3. Be sure all screws and connections are tight, and that the tension on the switch is adequate to assure good contact.

4. Recheck all connections against the figure.

The telegraph set is now complete and ready to be tested.

TESTING (figure 22)

Move the send/receive switch to the send (open) position and connect the dry-cell battery. Press the key; check that the metal sounder strip pulls down to the electromagnet with an audible click. The strip may have to be slightly bent to assure a good solid action. When the key is released the strip should spring up, causing another click when it hits the nail. Again, readjustment may be necessary.

FIGURE 22

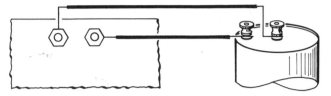

*Hookup for testing
the complete
telegraph set.*

Once the sounder is operating properly, the set may be connected for communications purposes (figure 23). The "line" wire should be #20 bell wire. If the two sets are installed between houses, requiring several hundred feet of wire, an extra #6 dry cell may be needed; if so, it can be hooked up as shown in the figure.

While using a two-way system, the receiver must keep his or her send/receive switch in the receive position and the sender must keep his or her switch in the send position. When the system is not in use, both switches should be left in the send position to prevent unnecessary drain from the batteries.

These sets should provide many hours of interesting and educational fun. They will operate for several months on one set of batteries.

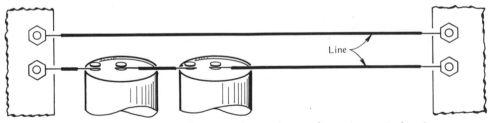

FIGURE 23   *Hookup for two stations. Don't forget to leave the send/receive switches in the "send" position when not using the stations.*

CHAPTER 4

"MR. WATSON, i WANT YOU..."

With these words, in June 1875, Alexander Graham Bell ushered in a unique development in electric communications. Now the human voice could be sent over a wire.

The operation of Bell's invention is based on the experiment we did in chapter 2 with the coils of wire, magnet, and compass. By moving a magnet near a coil of wire we produced a current flow in the wire, and a current flowing through another coil of wire caused a compass needle to move. Magnetic energy and electric energy can be readily converted into one another—which is what happened in the original telephone.

Figure 24 shows Bell's historic hookup. Each instrument has a coil of wire wound around a magnet and a disk of soft iron placed very close to the magnet but not touching it. The very thin disk is supported so that sound waves hitting its surface will cause it to vibrate, the way the bottom of a paper cup vibrates when you speak directly into it.

We also see in the figure that the magnetic energy (or field) extends through the disks and around the coils of wire. If nothing moves, there is

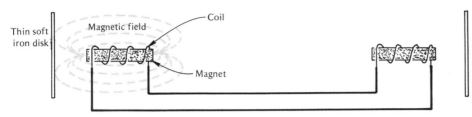

FIGURE 24 *Details of the first telephone.*

no changing field and nothing happens. But when someone speaks at one of the disks, it begins to vibrate and this movement causes the magnetic field to vary in step with the vibration. An electric current is produced in the coil of wire; it too changes in step with the sound. The varying current flows to the receiving instrument, where the coil of wire acts like an electromagnet and varies the pull on its own disk as the current gets stronger or weaker. The vibrations set up in the receiving disk match those in the transmitting disk and the original sound is reproduced.

The amount of current produced by this system is quite small. To make his telephone work, Bell had to shout into the unit while the listener pressed his ear against the receiver to hear the faint sounds.

You can duplicate Bell's telephone easily. Get a pair of magnetic earphones, the kind often used by electronic experimenters. Readily available from electronics hobby stores, these cost only a few dollars. They should have 1000–2000 ohms of impedance and screw-on ear caps (see the typical pair in figure 25). These earphones can be used for other projects later on.

When you get your "phones," unscrew the ear caps and slide the thin metal disk (called a diaphragm) off the magnet ends. The construction is the same as in figure 24. (Two coils wound on magnets offer increased sensitivity, but single-coil units are fine.)

To modify the earphones, unclip the headband. Determine how the wires coming into each earphone are connected, and disconnect them. Reconnect two 12-inch lengths of plastic-covered bell wire to the connections. Replace the diaphragm and screw the cap back on. When you are finished, you will have the two separate units (right side, figure 25).

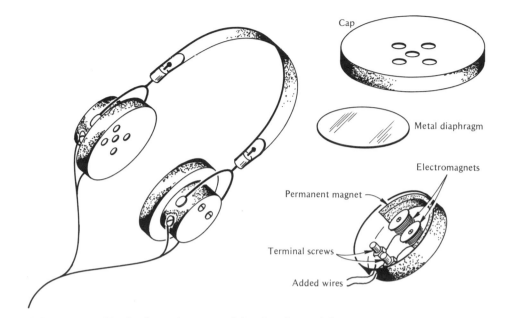

Cap

Metal diaphragm

Electromagnets

Permanent magnet

Terminal screws

Added wires

FIGURE 25   *Typical earphones and the simple modification necessary for experimentation.*

Each earphone is, in principle, a duplicate of Bell's telephone. Connect 10 feet of wire between each earphone unit (figure 26). Using rubber cement, attach a paper cup, with the bottom removed, to each unit. This will help concentrate the sound waves and make the telephone a little more sensitive.

Speak into one earphone unit and a friend will be able to hear you clearly in the other. Like Alexander Graham Bell, you may have to speak quite loudly!

You can build a telephone system that far surpasses the 1875 version, using a different voice-to-current transmitter. The current-to-voice receiver is still an earphone. The cup is not used this time.

Figure 27 shows a sensitive transmitting device, a microphone, that works like the one in the modern telephone. As in the original telephone, there is a thin diaphragm which vibrates when you speak at it. Instead of varying a magnetic field, however, it varies the pressure on a loose pile of carbon particles. Such particles are normally poor conductors of electric

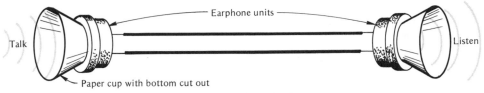

Earphone units

Talk

Listen

Paper cup with bottom cut out

FIGURE 26   *How to hook up an "earphone telephone."*

current, but when pressed together they become better conductors. If such a device is put in a circuit with a battery (to supply a substantial amount of current) and a receiver such as our earphone, the varying voice current is many times as strong as in the all-magnetic system. The circuit shown in figure 28 will reproduce communications loud and clear over hundreds of feet of wire.

Here is how to build a carbon microphone.

1. Cut and drill three pieces of pine or other wood (figure 29). The use of inexpensive hole cutters (available in most hardware stores) will make the job of cutting the large outer holes much easier. When drilling the third piece, be sure not to drill completely through. Remove any plug by prying with a screwdriver if necessary.

2. Obtain an old, used #6 dry cell. A good source is burglar alarm companies, which have old units that they discard. Carefully cut the case

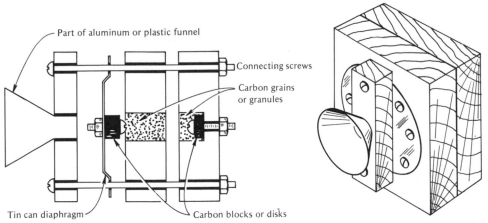

Part of aluminum or plastic funnel

Connecting screws

Carbon grains or granules

Tin can diaphragm

Carbon blocks or disks

FIGURE 27   *Carbon microphone that can be built by the experimenter (expanded view).*

FIGURE 28

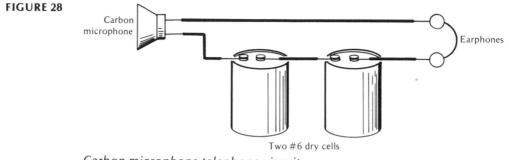

Carbon microphone telephone circuit.

of the cell lengthwise a third of the way (no farther) with a hacksaw. Pry off the rest of the case with a screwdriver and discard all the chemicals and other material; keep the central carbon rod.

3. Clean the carbon rod with water and a rag until all traces of chemicals are gone.

4. Make a diaphragm 3 inches in diameter out of the metal top from a 1-pound coffee can or 1-quart juice can. Drill a $\frac{3}{16}$-inch hole as close to the center of this top as possible.

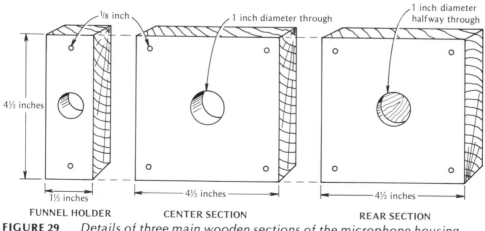

FIGURE 29   *Details of three main wooden sections of the microphone housing.*

FIGURE 30

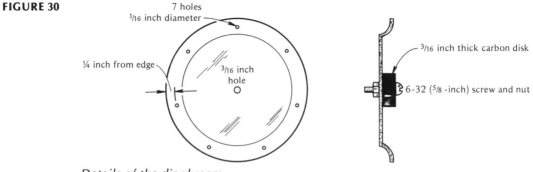

7 holes
¹⁄₁₆ inch diameter

¼ inch from edge

³⁄₁₆ inch hole

³⁄₁₆ inch thick carbon disk

6-32 (⁵⁄₈ -inch) screw and nut

Details of the diaphragm.

5. Drill eight holes, each ¹⁄₁₆ inch in diameter, around the rim of the top, ¼ inch in from the edge.

6. Cut two disks from the carbon rod, one ¼ inch thick and the other ³⁄₈ inch thick. Drill a ³⁄₁₆-inch hole in the center of each. Mount the ³⁄₈-inch disk on the diaphragm with a 6–32 (⁵⁄₈- or ¾-inch) screw. The diaphragm will look like figure 30.

7. Cut off a 1-inch length of carbon rod and wrap it in a piece of cloth. Hit the package with a hammer, breaking it into small chunks or granules about ¹⁄₁₆ inch in diameter.

8. Mount the ¼-inch-thick carbon rod disk in the rear of the appro-

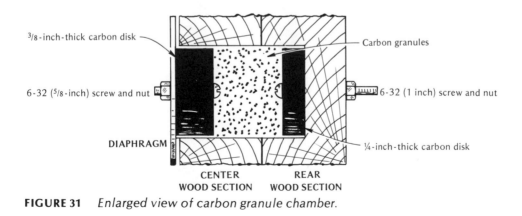

³⁄₈-inch-thick carbon disk

Carbon granules

6-32 (⁵⁄₈-inch) screw and nut

6-32 (1 inch) screw and nut

DIAPHRAGM

¼-inch-thick carbon disk

CENTER WOOD SECTION

REAR WOOD SECTION

FIGURE 31   *Enlarged view of carbon granule chamber.*

priate piece of wood (figure 31). Fill half the space remaining in the hole with carbon granules.

9. Line up the diaphragm over the carbon granules so that its carbon disk is free to move into the hole without binding. This is very important; any binding will prevent proper operation.

10. Screw on the diaphragm with seven #4 (¼-inch) round-headed wood screws. Tighten the screws securely.

11. Drill the eighth hole in the diaphragm through the wood base with a ³⁄₁₆-inch drill.

12. Add a 6–32 (2-inch) screw and nut in this hole as the diaphragm connection.

13. Finish assembly by adding the front piece of wood and a small funnel, if available. This funnel will help direct sound waves to the diaphragm.

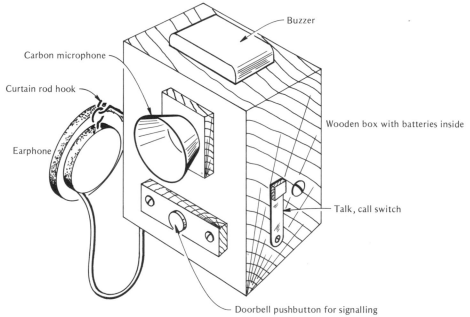

FIGURE 32 *Drawing of a complete telephone instrument that can be easily duplicated by the experimenter.*

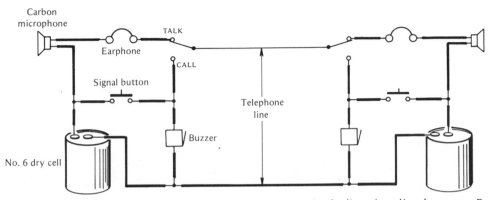
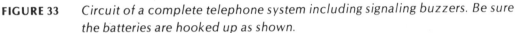

FIGURE 33   *Circuit of a complete telephone system including signaling buzzers. Be sure the batteries are hooked up as shown.*

To test the completed microphone, connect it to an earphone and dry cell (figure 28), hold it upright, and speak at the diaphragm. You should hear your voice clearly in the earphones.

Figure 32 shows a complete telephone set built with microphones obtained from old telephones. You can use the microphone you just built; however, commercial units are often more sensitive and offer greater clarity. Often, old microphones can be had for the asking from your local telephone company service department or a private telephone repair shop. Be sure to get a carbon-type microphone.

Figure 33 is a schematic diagram of the complete two-station telephone hookup, including buttons and buzzers for signaling and a call/talk switch. The switch, which is like the send/receive switch used in our telegraph, should be left in the call position except when you are actually talking.

This telephone station should operate at distances of up to 1000 feet if properly constructed. It will provide many hours of fun—not to mention savings on phone bills.

CHAPTER 5

Amplification and Long-distance Circuitry

The basic telegraph and telephone sets we have built work well over short distances. Commercial equipment, however, must operate over hundreds and even thousands of miles, and there had to be developed methods of increasing the strength of the telegraph or voice currents to levels that can span such long distances. At the receiving end, the necessarily weak current coming in has to be increased to a level that will operate a sounder or earphone.

Mechanical amplifiers—relays (figure 34)—were used to accomplish this amplification at first. The relay is similar to the telegraph sounder of chapter 3, but the steel strip, when attracted by the magnet, does not move far enough to make an audible click. Instead, it moves far enough to complete another circuit, which in turn causes the sounder to click. The short leads in this local circuit provide plenty of current for a loud, clear signal. Such relays were widely used in the early telegraph days; they operated with signals that had traveled over many miles of wire.

FIGURE 34

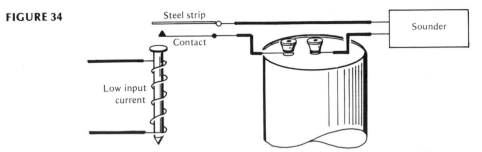

Operating circuit of a relay — one type of mechanical amplifier.

A different type of device is necessary to strengthen voice currents. It must smoothly amplify the rapidly moving currents that correspond to speech, without changing them. Changes can cause unrecognizable or distorted output sounds.

The first device to properly amplify voice currents was the vacuum triode (vacuum tube) developed by Lee De Forest in 1907. Like the relay, this uses a second circuit in which strengthened currents are produced; but here the similarity ends.

Figure 35 shows a vacuum triode in a typical amplifying circuit. The tube consists of a glass envelope containing a filament (like the one in a light bulb), a metal plate, and a screen or thin wire grid. The air has been pumped out, and the glass envelope sealed. A battery heats the filament until it glows. Another, higher-voltage battery is connected in a circuit that consists of an earphone, one side of the filament, and the metal plate in the triode. Because electrons will flow through the vacuum from the heated filament to the metal plate, a current flows in the earphone portion of the circuit.

Directly in the path of the flowing electrons, however, is the grid. As De Forest found, when very weak signals are applied between grid and filament, the flowing electron stream is varied by the grid in exact step with the incoming signals. The current in the plate circuit is many times as large as the weak grid current, so an amplified version flows through the earphone.

FIGURE 35

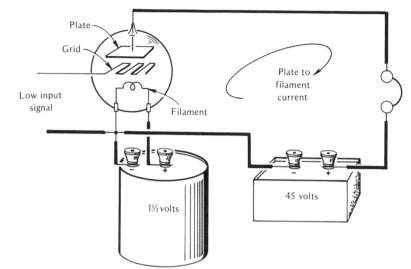

Plate

Grid

Low input signal

Filament

Plate to filament current

1½ volts

45 volts

De Forest's vacuum triode—the device that made the electronics industry possible.

In De Forest's vacuum tube, amplifications of four to five times could be achieved. Modern tubes can amplify hundreds, even thousands of times. Without the vacuum tube, the development of the electronics industry—radio, television, radar, and other kinds of electronics communications—would have been impossible.

In 1948 the transistor, another amplifying device, which is gradually replacing the vacuum tube in almost all applications, was developed by scientists at the Bell Telephone Laboratories in New Jersey.

A typical transistor (figure 36) consists of a tiny bar of silicon or germanium (a rare metal) which has been specially treated so that three distinct regions, the emitter, the base, and the collector, have been formed. Without going into all the details, we can say that the transistor has functions very similar to those of the vacuum tube. The current, its strength controlled by the signal applied to the base, normally flows between emitter and collector. Since collector current is many times as high as base current, amplification occurs. The big advantage over vacuum tubes is that filament batteries and high voltages are unnecessary: most transis-

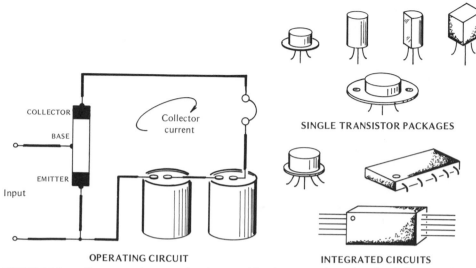

COLLECTOR

BASE

EMITTER

Input

Collector
current

OPERATING CIRCUIT

SINGLE TRANSISTOR PACKAGES

INTEGRATED CIRCUITS

FIGURE 36 *Construction, packaging, and basic operating circuit of the transistor.*

tor circuits operate at less than 12 volts. Transistors, as is well known, are much smaller than vacuum tubes; the actual transistor element is so small that you would need a microscope to see it. Hundreds, even thousands of these devices are often packaged into small modules known as integrated circuits, which are complete functional building blocks such as amplifiers, radio and TV processing circuitry, and computer logic elements. No matter how complex the integrated circuit is, it is still made up of individual components—but very tiny ones.

To see how a transistor works, we will need the two earphones used in chapter 4, two 1½-volt dry cells, a 1000-ohm variable resistor, and a general-purpose NPN transistor, type 2N4123 or similar type. The last two items may be obtained at the electronics hobby store where you bought the earphones and should cost about $1.50.

Connect the earphones together as you did when building the simple telephone. Have a friend talk (or shout) into one while you notice how low the sound level is from the other.

Connect the earphones to the transistor variable resistor and bat-

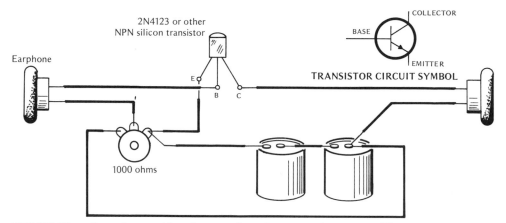

FIGURE 37   *Experimental hookup to show how transistors amplify.*

teries (figure 37). Be sure that all connections are correct. Again have your friend speak into one earphone while you adjust the variable resistor and listen to the other one. When you reach the correct transistor input voltage, the voice signal will be many times as loud as before. A variable resistor is necessary, since transistors must have a minimum voltage at the base in order to operate. In commercial practice, amplifiers are made with many transistors or integrated circuits and can raise the levels of signals thousands of times.

We will use amplifiers later, but it is not necessary to connect amplifying devices to the telephone or telegraph systems we have built, since the range of our equipment is a few hundred feet at most. If reception is weak because your line is a little long, adding one or two batteries will usually solve the problem.

There remains the problem of all that wire. A 200-foot run requires 400 feet of wire. A simple way around this is to use the ground as one conductor. You can see that the earth conducts electricity by connecting the conductivity tester of figure 6 to two 18-inch metal rods made out of an old wire coat hanger. Scrape any insulating paint or varnish off the coat hanger with sandpaper. Straighten the rods carefully and push them into normal soil, a foot or so apart, until only 2–3 inches are exposed. Connect

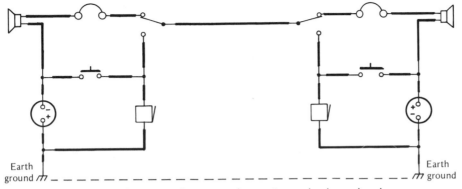

Earth ground

Earth ground

FIGURE 38   *Using the ground as a conductor in a telephone hookup.*

the tester to the two rods and the lamp should light. If it does not, it is because the soil is very dry. A bucket of water between the rods will solve the problem.

To use the ground as a conductor in the telephone hookup of chapter 5, look at the circuit of figure 38. The telephone hookup is wired so that one wire goes to the ground. A good connection to the ground must be made: the best way is to find a cold-water pipe close to where you are going to set up your telephone. Clean a short section of the pipe with emery cloth and sandpaper. Attach the wire by winding it five to ten times around the pipe and tightly taping the entire connection. Use strong electric or cloth tape. Since cold-water pipes connect directly to the water main buried under the street in front of your house, there are many feet of connecting area to the ground.

You may have to increase the number of dry-cell batteries for the buzzers to work if you use this method, but with good ground connections no more than three #6 dry cells or one 6-volt lantern cell on each end should be necessary.

CHAPTER 6

TALKING ON A LIGHT BEAM

Sooner or later the experimenter gets tired of the constraints of wires in his communications systems. Even installing a single wire and ground return can be a serious problem. For a "wireless" system, there are two alternatives. One, of course, is radio transmission, which requires a government-issued license and strict observance of rules and regulations. The other, which has no such complications, is the use of a common beam of light to convey voice or telegraphic signals.

Once you have learned the international Morse code, an elementary two-way light-beam system can be made of an old flashlight, a few pieces of wood, some batteries, our homemade telegraph key, and a simple lens. These components will go into the transmitter (figure 39) and our eyes will be the receiver.

The transmitter's light source is the reflector and bulb from a discarded flashlight, mounted in a hole bored through a piece of wood. Not far in front of this reflector, in a similar wooden holder, is a simple convex (magnifying) lens of the type sold in toy shops.

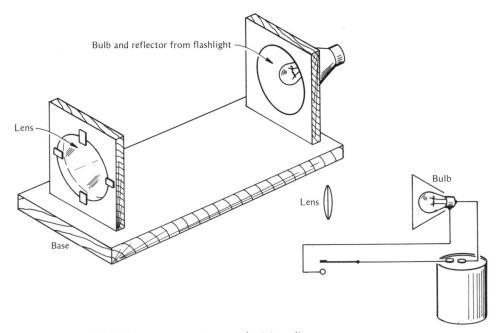

Bulb and reflector from flashlight

Lens

Base

Lens

Bulb

FIGURE 39   *Light-beam transmitter and wiring diagram.*

The lens collimates (concentrates) the light from the lamp in a narrow beam. (Without the lens, the light would spread out and become too weak to see clearly at the receiving end.) Figure 40 shows what the lens does: note how changing the distance from bulb to lens changes the output beam.

To build the transmitter (figure 41), align the lens by a simple procedure. In a dark room, or at night, set up the system so that it will project its beam of light on a wall or other "screen" about 20 feet away. Adjust the lens spacing so that a magnified image of the bulb is formed on the screen. This assures maximum concentration of light.

To send two-way messages, build two transmitters and arrange them so that there is a clear path between the two stations. With careful alignment, hundreds, even thousands of feet can be covered. A pair of inexpensive binoculars or a telescope will further increase the receiving range.

(A) Bulb far away from lens, light focuses at a point.

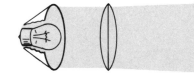

(B) Bulb too close to lens, light diverges.

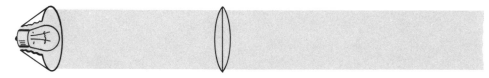

(C) Bulb at correct spacing from lens, light is collimated (parallel).

FIGURE 40 *The effect of lens position on the output beam of light. The reflector adds light coming from the rear and sides of the bulb to the forward beam.*

It is not necessary to have the batteries and lamp assembly in the same room; it may be a good idea to mount the lamp in a high clear spot on a roof or upstairs window, and the batteries and key in a more accessible location. Be sure you can see the other station's transmitter clearly. If you decide to mount the transmitter outside, provide protection from the weather.

By adding components you can eliminate the need to view the transmitter directly, setting the stage for transmitting voice signals over the light beam.

Figure 42 shows a two-way station that can send telegraph signals more than 1000 feet when the stations are mounted in the attics of two

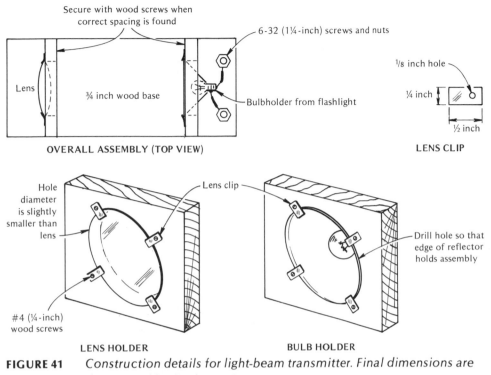

Secure with wood screws when
correct spacing is found

6-32 (1¼-inch) screws and nuts

Lens

¾ inch wood base

Bulbholder from flashlight

OVERALL ASSEMBLY (TOP VIEW)

⅛ inch hole

¼ inch

½ inch

LENS CLIP

Hole
diameter
is slightly
smaller than
lens

Lens clip

Drill hole so that
edge of reflector
holds assembly

#4 (¼-inch)
wood screws

LENS HOLDER

BULB HOLDER

FIGURE 41 *Construction details for light-beam transmitter. Final dimensions are determined by final part sizes.*

houses facing each other. The transmitter has been modified (figure 43) to use a brighter lamp, extra batteries, and a doorbell buzzer to produce an interrupted light beam. Be sure that you hook up to the correct point of the buzzer contact. If the light beam is not interrupted, you will hear only one click when the light is turned on, and one more when it goes off. At a distance, with a weak signal, these clicks could be missed, making reliable communication difficult.

The big change is in the receiver. Since we are not going to view the transmitted light directly, we must use a device, the photodetector or photocell, that will detect it. Photodetectors, which come in many types, convert light into an electric change—in resistance (when exposed to

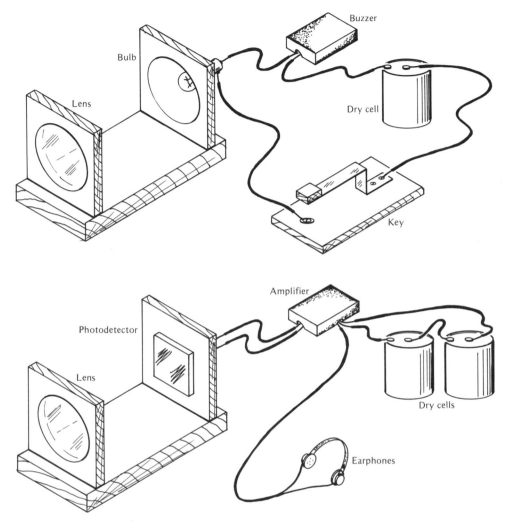

FIGURE 42 *Complete two-way light-beam station. (See figures 43 and 47 for hookup details.)*

light), in amperage (by generating current), or in voltage (like a battery). We have chosen a voltage generator, which is used to produce electricity from sunlight and is sold by many electric supply shops as a silicon solar cell or battery.

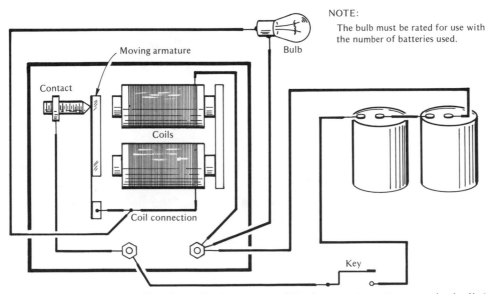

NOTE:
The bulb must be rated for use with the number of batteries used.

FIGURE 43 *Electrical hookup of a buzzer-modified transmitter. Be sure the bulb is connected across the buzzer coils. It* must *flicker to work.*

The voltage generator (figure 44), is about ½ x ½ inches and works as a detector for the transmitted signal. The voltage produced is very small, particularly when the light source is far away, so a receiving lens and amplifier must be used.

FIGURE 44

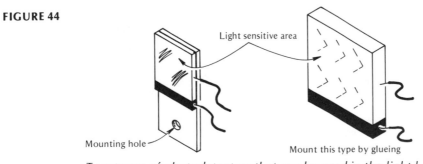

Two types of photodetectors that can be used in the light-beam receiver.

FIGURE 45 *Hookup for testing a photodetector.*

You can perform an easy experiment to see how the detector works. Buy a silicon solar cell: cost should not exceed a couple of dollars. Connect it to a pair of earphones (figure 45) and place it in front of the modified light-beam transmitter. When you press the key you can hear the buzz from the transmitter through the earphones, even a few feet away. The lamp is going on and off too fast for you to see it. The photocell "sees" it, however, and produces a voltage that also goes on and off. The earphones in the circuit then convert the varying voltage to sound.

In the receiver (figure 46), the photocell is mounted in a wood frame at the focus of as large a lens (preferably glass) as you can find. (A large lens collects more light from the transmitter.) You may find such lenses in powerful magnifying glasses, at opticians who manufacture eyeglasses, or at novelty shops and toy stores. There is one toy manufactured for children 1–3 years old, a 6-inch magnifying lens in a wooden holder, that is perfect for use in the receiver.

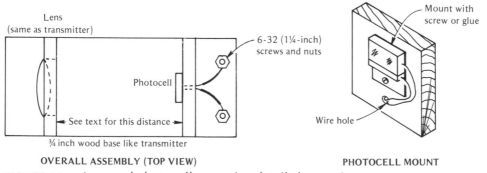

OVERALL ASSEMBLY (TOP VIEW)   **PHOTOCELL MOUNT**

FIGURE 46 *Lens and photocell mounting details for receiver.*

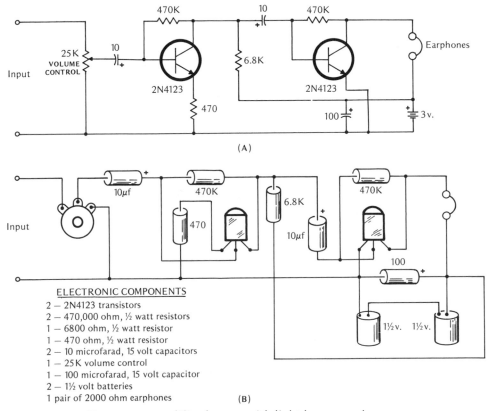

470K

10

470K

25K
VOLUME
CONTROL

10

Input

6.8K

Earphones

2N4123

470

2N4123

100

3 v.

(A)

10μf

470K

6.8K

470K

Input

470

10μf

100

1½ v.   **1½ v.**

ELECTRONIC COMPONENTS

2 — 2N4123 transistors
2 — 470,000 ohm, ½ watt resistors
1 — 6800 ohm, ½ watt resistor
1 — 470 ohm, ½ watt resistor
2 — 10 microfarad, 15 volt capacitors
1 — 25K volume control
1 — 100 microfarad, 15 volt capacitor
2 — 1½ volt batteries
1 pair of 2000 ohm earphones

(B)

FIGURE 47   *Two-stage amplifier for use with light-beam receiver.*

When the lens is mounted, align it like the lens we used in the transmitter. Remove the shade from a household lamp and place it 10–20 feet (or more) from the lens-photocell assembly. Turn off all lights except the lamp and carefully move the lens and photocell until a small, clear image of the light bulb can be seen at the center of the photocell. Fix everything in place, checking that the tiny image remains centered.

Now build the two-stage transistor amplifier in figure 47. The components can be obtained for a few dollars from an electronics supply store. Connected together, they form a circuit that amplifies the very

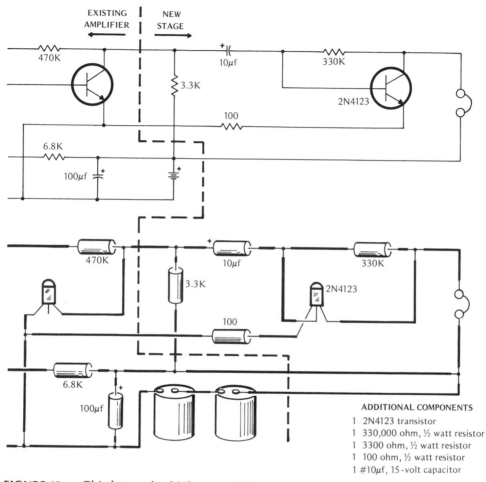

FIGURE 48   *Third stage for higher sensitivity.*

ADDITIONAL COMPONENTS

1 2N4123 transistor
1 330,000 ohm, ½ watt resistor
1 3300 ohm, ½ watt resistor
1 100 ohm, ½ watt resistor
1 #10μf, 15-volt capacitor

small voltages from the photocell to levels that will provide comfortable volume from a pair of earphones. A third stage for even more amplification may be added (figure 48), but may prove unstable.

If possible, solder the components of amplifiers together instead of clamping them with screws; soldering reduces the number of poor connections. The assembly drawings show this and suggest the use of brass

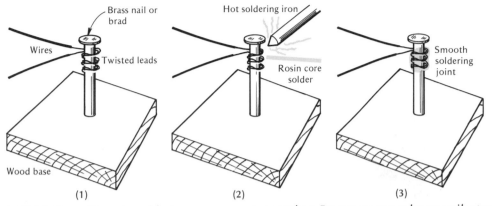

FIGURE 49 *Steps in soldering components together. Be sure to use a brass nail; steel or aluminum will not work.*

nails as connection points. Wrap the wires around the nail (figure 49). Heat the nail and wires with a pencil-type soldering iron (30–50 watts). A higher-wattage iron is unnecessary. After 4–5 seconds touch the heated joint with 60–40 rosin core solder until a small quantity of solder melts and flows over all the points to be connected. This should not take more than 5 seconds. Do not overdo it; use solder sparingly. Beginners often use too much solder or don't allow enough time to heat the connections properly. Try to get an experienced solderer to help you learn; with correct guidance it won't take more than 15 minutes. If you plan extensive electric experimentation, soldering is a skill you must learn. NEVER USE SOLDER OTHER THAN ROSIN CORE. IT WILL CORRODE AND RUIN YOUR PROJECT!

After double-checking the wiring, make final adjustments and tests. Set up a transmitter 10–15 feet away from the receiver and have someone press the key. If you can clearly hear the buzz when the transmitted beam of light shines on the photocell, everything is working. Increase the distance to 50 feet and try again. As the distance increases, alignment becomes more critical. Slightly readjust the lenses on transmitter and receiver for whatever improvement is possible. Ultimate range with lots of patience and a three-stage amplifier is about 1000 feet. It sometimes

FIGURE 50

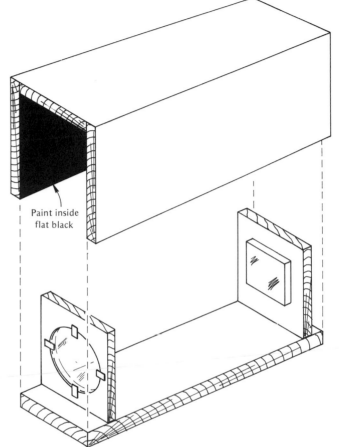

Light cover that can be used to shield the receiver.

Paint inside flat black

takes several hours to achieve this, however, so don't get discouraged too quickly.

If the light-beam equipment is to be used in daylight, it is important that the receiver only receive light from the transmitter. The lens-to-photocell path must be shielded so that stray light does not interfere. Add a short light baffle (figure 50). Larger-diameter lenses, particularly on the receiver, will help increase the strength of the received light beam and minimize the effects of stray light.

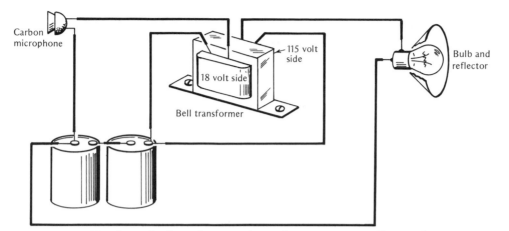

FIGURE 51   *Hookup of microphone and transformer to transmit speech.*

It may have occurred to you that if the interrupted light beam can be made to flicker in step with speech, we can send sounds over our system. This is possible and we will proceed to do it. The receiver requires no modification. For the transmitter, we need two more items: a carbon microphone of the type built in chapter 4 and a common doorbell transformer, available from most hardware stores for $1–$2. The transformer couples the microphone to the lamp; it should have a 10–18-volt secondary and a 115-volt primary. The function of a transformer is to change one voltage level to another. In our circuit, it changes the microphone voltage to the proper level to operate the lamp. The new transmitter hookup is shown in figure 51.

When the wiring is complete, connect the batteries and speak into the microphone. The lamp should flicker as you talk, indicating that the speech voltages are modulating the lamp. If the flickering is not very obvious, try adding an additional battery; be careful not to burn out the lamp, however.

The transmitter can be placed in line with a receiver and the system checked out. Clear sounds should be easily sent. In figure 52 the voice and telegraph transmitters have been combined into one and the user can communicate by either method depending on the position of the switch.

FIGURE 52

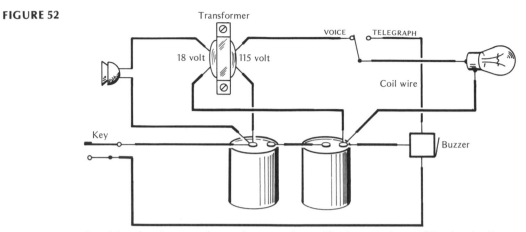

Transformer

18 volt   115 volt

VOICE   TELEGRAPH

Coil wire

Key

Buzzer

Combined voice or telegraph transmitter. The buzzer is modified as in figure 43.

FIGURE 53

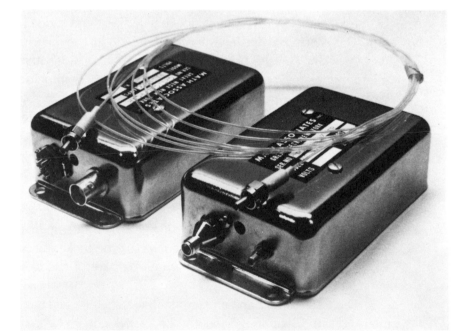

A typical fiber-optic transmitter, receiver, and fiber capable of transmitting a wide range of electrical signals.

In a variation of light-beam communications, thin glass fibers have been developed which have the uncanny ability to conduct light over long distances. These optical fibers have been developed to the point where they will only lose a small amount of light after thousands of feet. Since they are unaffected by moisture, changes in temperature, and most corrosive substances, their use in industry will increase.

Specially designed units used as light sources in fiber-optic systems respond not only to voice signals, but to television and computer signals. Figure 53 shows a typical fiber-optic data link and a short length of glass fiber; note the connectors, which look like electrical plugs, and the plastic protective jacket on the glass fibers. As in our equipment, one of these "modules" converts electric signals into varying light, which travels through the fibers to a receiving module, where it is converted back into electric signals that match the original ones.

Stockton Township Public Library

, Stockton, Illinois

CHAPTER 7

Radio

When radio broadcasting was in its infancy, young people, excited by the magic of wireless communications, built all sorts of transmitting and receiving devices, which enabled them to learn much about the field which some of them later made their profession. Today mass-produced, inexpensive radios are readily available, but the educational value of hearing wireless signals on homemade equipment remains.

Figure 54 illustrates a simple experiment on the transmission of information over a short distance with no direct connection, using two large coils of wire and a buzzer. It operates on principles we have already investigated. When the key is depressed, pulsing electric current flows through the transmitting coil, creating a pulsating magnetic field outside the coil. When the magnetic field passes through the receiving coil, it is changed back into an electric current (also pulsating), and we hear the original buzzing in the earphones. (This fact was proved in the compass needle experiment in chapter 2.) If the earphones are sensitive and two or

FIGURE 54

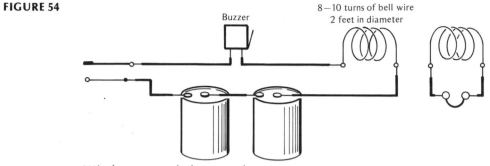

Buzzer

8—10 turns of bell wire
2 feet in diameter

Wireless transmission experiment.

three batteries are used to power the buzzer, the coils will transmit over surprising distances. Early wireless experimenters used coils as much as 10 feet wide and achieved reasonably long-range communications.

Whenever a magnetic field is made to vary rapidly, an electric field or radio wave is also produced. Radio waves vary at a fixed rate (or frequency) and they can travel great distances through the air. The path through the air is much like a common electric circuit (see figure 55). One "wire" may be thought of as the air, and the other as the ground.

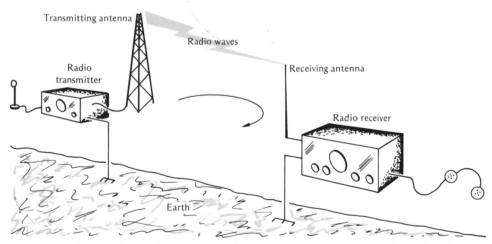

Transmitting antenna

Radio waves

Receiving antenna

Radio transmitter

Radio receiver

Earth

FIGURE 55 *Comparison of radio transmission with a common electrical circuit.*

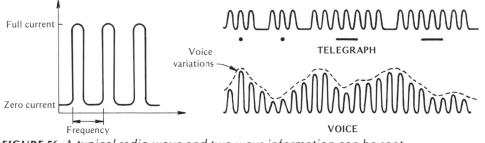

FIGURE 56 *A typical radio wave and two ways information can be sent.*

The radio waves produced by our coils are so weak that they are of no practical value. They are true waves, nevertheless, produced the same way as their commercial counterparts.

Figure 56 is a graph of the current that flows through the buzzer/transmitting coil circuit. When the buzzer contacts are closed, current flows; when they are open, it does not. This on-off or interrupted flow is what will produce the radio wave. The strength of the wave is a function of the strength of the current flowing and the frequency of the wave (a function of the frequency or tone of the buzzer).

With several buzzers we can produce radio waves of different frequencies. Commercial transmitters do not of course use buzzers but special vacuum tubes or semiconductor circuits. Information is sent as in the light-beam transmitter of chapter 6. The stream of waves is turned on or off for telegraph signals, or is varied in strength for voice transmissions (figure 56). The radio wave can also be varied in frequency.

We will examine receivers of signals from commercial radio stations; then we will produce our own signals.

To receive radio waves we need a device, called an antenna, to "capture" the waves. It is a long wire placed high up and out in the open, and insulated so that the waves it "catches" are only allowed to go where we want them to (figure 57). Materials for putting up an antenna include some ordinary bell wire, a few hardware store hooks, and insulators—which we can make from any nonconducting material, but for best re-

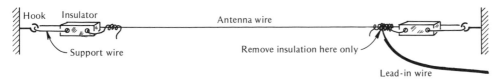

FIGURE 57 *Construction and parts of an antenna.*

sults nonporous ceramics or plastics should be used. Figure 58 shows several suitable insulators, including commercial ones (if they can be found) that will work fine.

To install the antenna, choose two high points such as trees, poles, or sides of buildings, as far apart as possible: 50–100 feet is perfect. Attach the insulators to the two support structures, string the wire, and run the lead to where you will set up the final station. Be sure that the lead-in wire is properly supported and remains insulated. Figure 59 shows a number of ways antennas can be installed. The longer and higher the installation, the better, but over 100 feet is wasted effort. If you can only set up

Glass "EGG" insulator

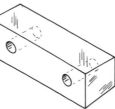

Drilled plastic block

Glass "STRAIN" insulator

Drugstore pill case with holes

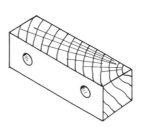

1 x 2 wood with several coats of varnish

FIGURE 58 *Common insulators that are suitable for building antennas.*

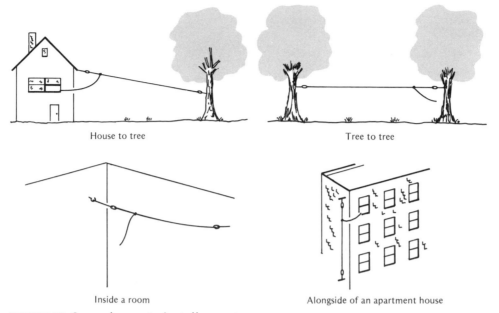

House to tree   Tree to tree

Inside a room   Alongside of an apartment house

FIGURE 59 *Several ways to install an antenna.*

20 feet of wire, try it—it may work. NEVER USE A POLE SUPPORTING ELECTRIC WIRES AS AN ANTENNA SUPPORT. DO NOT STRING THE ANTENNA SO IT CAN FALL ON ELECTRIC POWER LINES OR SO THEY CAN FALL ON IT. Every year experimenters—some quite knowledgeable—are seriously injured or even killed by not observing this simple precaution.

After the antenna is installed, work on the other end of the circuit, the ground connection. As in chapter 5, we can use a cold-water pipe; or a good ground connection can be made from a 3–4-foot-long iron or steel pipe. The pipe should be hammered into the ground and a wire run from the exposed end to the station. Be certain that the connection to the pipe is well made: scrape the pipe with sandpaper and tightly twist the wire in place.

Next comes the receiver. See figure 60 for the most elementary receiver. It requires only our 1000-ohm earphones and a specially made

FIGURE 60

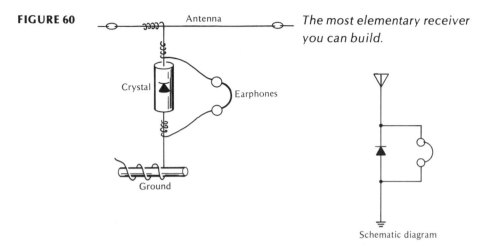

The most elementary receiver you can build.

Schematic diagram

detector consisting of a tiny crystal of germanium or silicon with hairlike wires attached to specially prepared portions of it. The crystal has the unique ability to convert radio energy into common, varying electric currents. In the early days of radio, chunks of iron ore or galena or similar crystals were mounted in holders and "probed" with thin wires to make detectors. Today sensitive germanium or silicon crystals are prepackaged, fully adjusted, and available from most electric supply shops (figure 61 indicates some of the common forms and part numbers that are readily available). Try to obtain germanium crystals, as they are most sensitive.

In the receiver of figure 60, radio waves are collected by the antenna and converted into varying electric currents by the detector; these flow

FIGURE 61

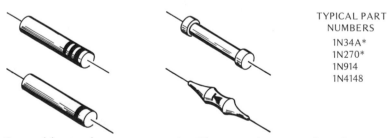

TYPICAL PART
NUMBERS
1N34A*
1N270*
1N914
1N4148

Several forms detectors can take. The starred ones are best for crystal detector use.

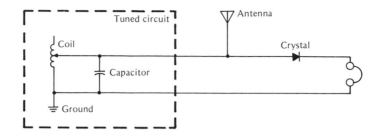

FIGURE 62   *Addition of a tuned circuit to the simple receiver of figure 60.*

through the earphones, allowing us to hear the information being transmitted. When you listen, you will almost always receive a number of different stations simultaneously. The detector cannot select from the signals passing through; it responds to them all.

To hear the stations you prefer, it is necessary to filter out the others with a coil of wire and a capacitor connected into a tuned circuit. Tuned circuits only allow one frequency to pass through them. In figure 62 the coil is adjustable so that we can select (tune in) different frequencies.

To make a tuned circuit, begin with the coil. Wind a 3-inch-wide layer of #26 enameled magnet wire on a 5-inch length of 2½-inch clam casing, a common lumberyard molding. The wire is obtainable in most electric supply stores and you can get the wood almost certainly as a present from any lumberyard—especially if you have been buying lumber from them for other projects. As you wind the coil be sure that all turns are right next to each other, neatly and tightly. "Paint" the coil with three coats of varnish, allowing each coat to dry completely before applying the next one. The slider can be fabricated of tin-can stock. When the coil is finished and mounted (figure 63), slide the movable contact over the turns, applying enough pressure so that the top layer of varnish is scratched in a path following the contact point of the slider. Take fine-grade sandpaper and sand the scratched path until bright copper shines through. If any turns loosen, revarnish the coil and start again. When the

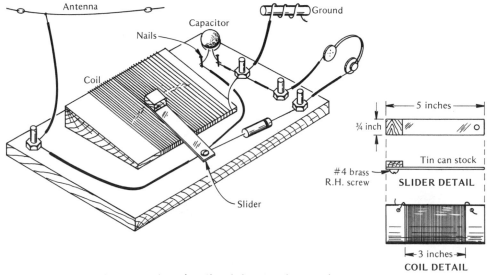

FIGURE 63   *Construction details of the simple tuned receiver.*

coil is properly prepared, good contact between each turn and the slider will occur as the slider is removed.

When the coil is completed, buy a mica or ceramic disk capacitor of 200–300 micromicrofarads (abbreviated mmf or pf) from an electronics component shop. It costs only a few cents. Hook everything up as in the figure and listen. Moving the slider should allow one, two, or more stations to be "peaked" or tuned in.

The radio signals picked up by this simple receiver are weak electrically, which is why all the stations in your particular area can't be tuned in. Try adding an amplifier (figure 64), much the same way an amplifier was added to the earphone telephone in chapter 4. Adding an amplifier to the sound-carrying section of the receiver will greatly increase the volume of received speech.

With or without an amplifier, the receiver is limited in overall quality of the tuned circuit. There have been many improvements since such circuits were first used, especially in the quality of the coils. Figure 65 shows two commercially available coils which are sold under such

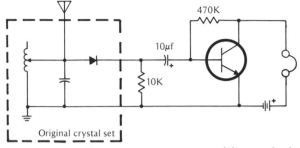

ELECTRONIC COMPONENTS

1 470,000 ohm, ½ watt resistor
1 10,000 ohm, ½ watt resistor
1 #10μf, 15-volt capacitor
1 2N4123 (or similar) transistor
2 1½ volt batteries

FIGURE 64 *How to connect an amplifier to the basic receiver.*

names as vari-loopstick, ferri-loopstick, and ferrite antenna coils. They are all wound on a rod of specially prepared iron particles, which increases the filtering action of the coil many times over the air-wound coil we built. If you purchase one, be sure to get the kind with the screwdriver-adjustable core, as shown in the figure. A matching knob is available that makes this adjustment easier. Putting it in the circuit in place of the air-wound coil and slider will greatly improve the receiver. You need only turn the screw to tune in different stations.

FIGURE 65

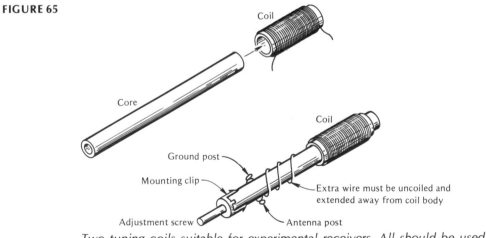

Two tuning coils suitable for experimental receivers. All should be used with a 200-300-picofarad capacitor for best results.

In the early days of radio, amplifying tubes were very expensive and transistors had not been invented, so ingenious ways were found to make one tube do the work of many. The popular regenerative receiver was, and still is, considered extremely sensitive, amplifying thousands of times, but it is also rather hard to tune. Building one will lead us into the next two chapters by demonstrating some important principles. The operation is so much better than that of a simple crystal set that it pays to build one for use as a personal receiver.

We will use a transistor rather than old-fashioned vacuum tubes. You should realize, however, that the receivers produced up to the 1950s used tubes.

The tuned circuit portion of the regenerative receiver (figure 66) is similar to the one in our basic receiver. The big change is adding an extra winding, called the tickler or feedback coil, to the main tuning coil. When a radio wave enters the circuit from the antenna, it is amplified first by the transistor. The amplified signal flows through the feedback coil, which is directly in the circuit of the collector of the transistor. Current flowing through a coil, as we know, produces a magnetic field. The magnetic field flows through the tuning coil and feeds back radio wave current to the transistor, where it is again amplified. This feedback happens

FIGURE 66

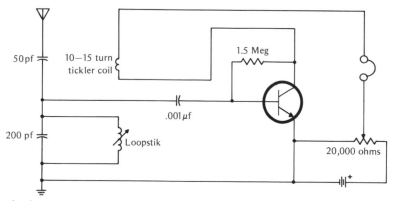

The basic regenerative receiver circuit. The 20,000-ohm variable resistor controls the transistor's operating point and thus the amount of feedback.

over and over until a point is reached where the whole circuit becomes unstable—or, as engineers say, begins to oscillate. Then all we hear through the earphone is a loud, uncontrolled screech. But by carefully reducing the amount of energy fed back we can find a point where we get maximum amplification with no screeching. This is the correct operating point.

Figure 67 shows a slightly different arrangement of parts in a regenerative receiver. The operation is the same. The feedback coil is wound

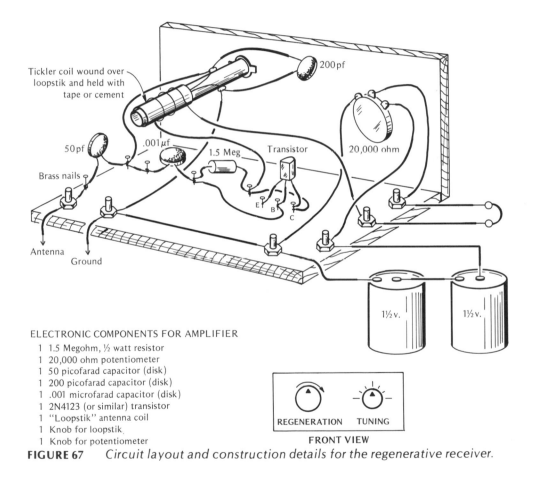

ELECTRONIC COMPONENTS FOR AMPLIFIER
1   1.5 Megohm, ½ watt resistor
1   20,000 ohm potentiometer
1   50 picofarad capacitor (disk)
1   200 picofarad capacitor (disk)
1   .001 microfarad capacitor (disk)
1   2N4123 (or similar) transistor
1   "Loopstik" antenna coil
1   Knob for loopstik
1   Knob for potentiometer

FIGURE 67   *Circuit layout and construction details for the regenerative receiver.*

directly over the winding of a commercial loopstick; the amount of feedback is controlled by a variable resistor which adjusts the amount of amplification of the circuit. The transistor also converts the amplified radio currents into sound currents, so we see that the transistor has plenty of work in this circuit.

When the receiver is finished, recheck all connections carefully, then hook up the antenna, ground, batteries, and earphone and turn the variable resistor completely counterclockwise, then slowly clockwise until the circuit just starts to oscillate. If there is no oscillation and the wiring is correct, reverse the connections to the feedback coil. Make a slight counterclockwise adjustment of the variable resistor so that the screech just stops. This is the most sensitive position. When tuning in another station you may have to readjust the feedback control for the greatest sensitivity. You can receive stations hundreds of miles away under good conditions. Use a good antenna and ground for best results.

CHAPTER 8

A VERSATILE SHORTWAVE RECEIVER

Now it is time to build a radio that will enable us to extend our listening range to the entire world. The receiver will let us hear many commercial stations outside the broadcast band, as well as many amateur radio operators (see chapter 10).

The receiver, a variation of the regenerative receiver we built in chapter 7, includes a sound amplifier stage for greater sensitivity. It tunes to a wide range of frequencies by altering the turns of the input tuned circuit.

The layout of the common frequencies used for the transmission of radio signals is quite large (figure 68). Frequencies are expressed in cycles per second, or hertz, after a physicist whose work laid a theoretical basis for wireless telegraphy and radio, Heinrich Hertz. The AM broadcast band in the U.S. extends from 500,000 to a little over 1,500,000 hertz (1.5 megahertz). Our radio will tune from just above this range to 20 megahertz, in the shortwave or HF region, which contains international broad-

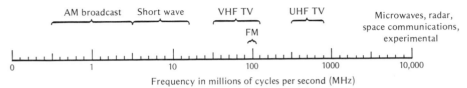

AM broadcast   Short wave   VHF TV   UHF TV   Microwaves, radar,
                                                              space communications,
                                   FM   experimental

0   1   10   100   1000   10,000

Frequency in millions of cycles per second (MHz)

FIGURE 68   *Layout of common frequencies used for transmission of signals. Note that the AM broadcast band is 1.5 MHz wide while the VHF TV band is 400 MHz wide.*

casters, police and fire departments, time stations, ship-to-shore, citizens radio (CB), and amateur (ham) radio. Covering such a large range with one tuning control would make selecting single stations too difficult; therefore, our receiver tunes this range in four smaller portions, or bands (see page 66). The bands are selected by selecting numbers of turns on the receiver's tuning coil and two variable capacitors—one for coarse adjustments, the other for fine adjustments.

A tuned circuit consists of a coil and a capacitor. We also saw in chapter 7 that if you vary the number of turns on the coil you can change the frequency the tuned circuit will respond to. Or you can change the frequency by changing the value of the capacitor. We will use variable capacitors in this receiver since they can be adjusted more finely, than a slider on a coil.

Figure 69 is a schematic diagram of the receiver. The basic circuit is regenerative; in this receiver, feedback occurs from the collector to emitter of the 2N5223 transistor by way of the 50 pf capacitor. To understand how the output of the transistor gets back to the input, recall the basic transistor circuit of figure 36, which shows how any current that flows through the collector also flows through the emitter. Since the collector is connected to the tuning coil, output energy then flows through the coil as well as the input signal and we have feedback. This arrangement assures the right connection for oscillation and eliminates the need to wind a separate coil.

The amount of feedback is controlled by a 500-ohm potentiometer whose setting determines how much voltage is applied to the 2N5223 and

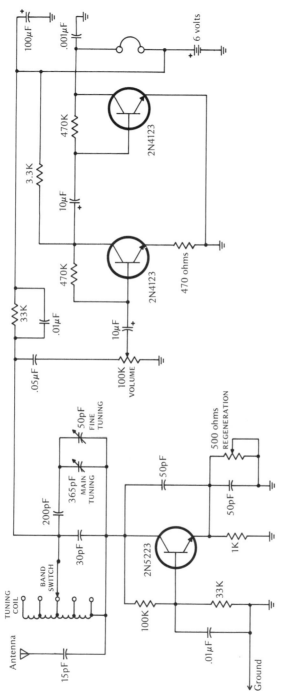

ELECTRONIC COMPONENTS

1 470 ohm, ½ watt carbon resistor
1 1000 ohm, ½ watt carbon resistor
1 3300 ohm, ½ watt carbon resistor
2 33K ohm, ½ watt resistors
1 100K ohm, ½ watt carbon resistor
2 470K, ½ watt carbon resistors
1 15pF disk capacitor
1 30pF disk capacitor
2 50pF disk capacitors
1 500 ohm linear potentiometer
1 100K ohm audio taper potentiometer
1 SP5 pos switch

1 200pF disk capacitor
1 .001μF disk capacitor
1 .01μF disk capacitor
1 .05μF disk capacitor
2 10μF electrolytic capacitors
1 100μF electrolytic capacitor
1 2N5223 transistor
2 2N4123 transistors
1 Broadcast radio variable capacitor (365pF)
1 50pF variable capacitor
Miscellaneous components

FIGURE 69 Complete schematic diagram of shortwave receiver.

therefore also determines its amplification factor. The 100K and 33K resistors connected to the base simply establish proper operating voltages. Conversion from radio energy to varying electric current takes place in the transistor without the need for a crystal detector. As a result, both radio energy and sound currents are present at the output of the transistor.

The .01 μf capacitor connected across the 33K resistor in the collector circuit forms a filter that blocks radio energy from reaching the sound-amplifying portion of the receiver. The second transistor is the sound or audio amplifier. Its two base resistors establish the proper operating voltage. A 0.001-microfarad capacitor in the final transistor keeps any radio energy that may have leaked through away from the earphones. The entire receiver works from 6 volts, which may be obtained from a lantern battery, D cells, or four #6 dry cells. Figure 70 is a pictorial

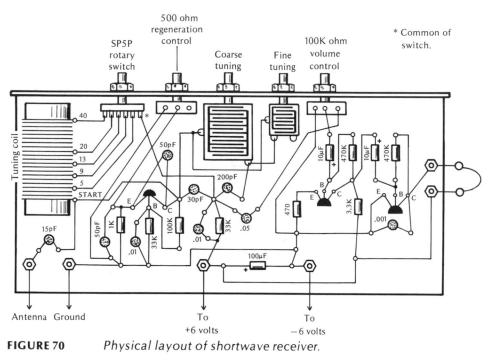

FIGURE 70   *Physical layout of shortwave receiver.*

layout of the receiver. Since it is a straightforward model and several sizes of parts can be used, we have left construction details to the reader.

As in chapter 6, it is important, in building this relatively complex, sensitive receiver, to make good connections when soldering the components to short brass brads, partly driven into the base as shown in the figure.

The parts for the receiver are commonly available from electronics supply shops, with the possible exception of the variable capacitors, which can often be salvaged from an old vacuum-tube radio; local repair shops are a good place to start looking. Another source for these capacitors is military surplus dealers, many of whom advertise regularly in electronics hobby magazines and often have catalogs of components that are indispensable to experimenters. If you purchase surplus parts of this type, check carefully for loose screws, dirt, or other foreign matter that

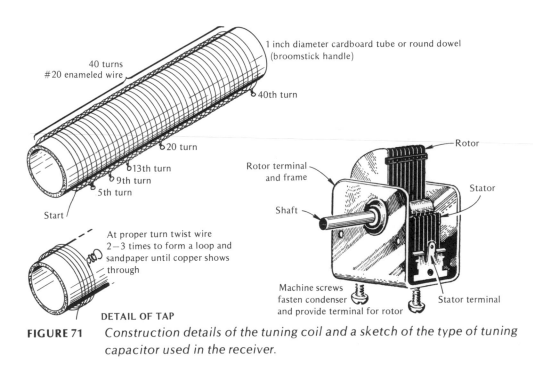

FIGURE 71 *Construction details of the tuning coil and a sketch of the type of tuning capacitor used in the receiver.*

can interfere with their operation. Clean them thoroughly. Exact mounting methods and dial knobs will be determined by the type of capacitor used. Figure 71 shows the details of making the tuning coil and the style of variable capacitor that may be used.

Check the wired receiver fully against the schematic and pictorial, carefully and slowly. Connect the antenna and ground leads and turn the tap switch to band 1. Set both capacitors and the volume control to midrange, set the regeneration potentiometer at its full counterclockwise position, and apply power. Slowly turn the regeneration control clockwise until you hear a hissing sound or oscillation start, then back off on the control slightly and turn the large variable capacitor back and forth. While setting the large capacitor, keep the small capacitor set at the center of its range. You should hear a number of chirps in the earphone, which are radio stations. Try to set the large capacitor as close to one of these as possible, and fine-tune it with the small capacitor. Adjust the regeneration control for best-sounding signals and you should hear the station clearly.

A couple of hours of practice with the three tuning controls should make you expert and you will be able to hear many stations of different types. Try changing the tap switch to other bands to find out what you can hear. The approximate coverage of this switch is as follows:

band 1   1.5 to 4 MHz (or megahertz)
band 2   4   to 10 MHz
band 3   9   to 15 MHz
band 4   13   to 20 MHz

Different frequency ranges are active at different times of the day, month, and year; an apparent "dead band" of one day may be the liveliest band the next day. Most international signals occur on bands 3 and 4, and the best time to listen for them is at night.

This shortwave receiver, while simple compared to commercially made equipment, will provide many hours of listening enjoyment.

CHAPTER 9

Radio Transmitters

Our experiments have reached the point where we need to progress to true wireless communications. Wires, light beams, and the like are too cumbersome and we want to use a true radio transmitter. This unfortunately poses a problem.

The U.S. government, through the Federal Communications Commission, has passed strict laws requiring that every radio transmitter of significant power and range be operated only by a licensed operator. The frequencies that can be used are also closely regulated, so you cannot go ahead and build transmitting equipment without careful consideration. Some licenses are not too difficult to obtain, and as a result, the hobby of amateur radio developed many years ago. (See chapter 10.) Without a license one is permitted to make only very low-power, short-range transmitters. Such transmitters will broadcast some hundreds of feet, however, and they permit communications to occur that would be impossible with the equipment built in previous chapters. Since the principles are the same as for higher-power transmitters, the experience is educational.

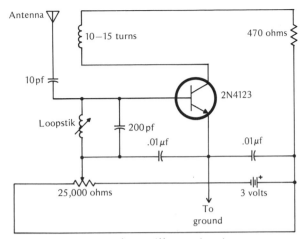

Antenna

10—15 turns   470 ohms

10pf

Loopstik

200 pf

2N4123

.01 µf   .01 µf

25,000 ohms   3 volts

To ground

ELECTRONIC COMPONENTS

1 Loopstik antenna coil
1 10 picofarad disk capacitor
2 .01 microfarad disk capacitors
1 200 picofarad disk capacitor
1 2N4123 (or similar) transistor
1 470 ohm, ½ watt carbon resistor
1 25,000 ohm potentiometer
2 1½ volt dry cells

FIGURE 72   *Simple oscillator circuit.*

A radio transmitter is basically a source of continuous radio waves that can be varied (modulated) to convey information. As we have already seen in several experiments, the radio wave carrier can be turned on and off to transmit telegraphic signals. It can be varied in intensity to transmit speech (amplitude modulation or AM) and it can be varied in frequency to transmit speech without static (FM). We must first produce radio waves, then modulate them.

Our circuit (figure 72) consists of a transistor and some components which make up an oscillator. Note that the regenerative receiver of figure 66 was an oscillator. The oscillator now being described contains a tuned circuit in the base lead and a variable resistor to supply the proper amount of voltage for operation, but, unlike the regenerative receiver, here the regenerative resistor is adjusted so that feedback always occurs. Output from the oscillator is taken through a small 10 pf capacitor.

The output is a single continuous wave at a frequency determined by the tuned circuit. To transmit this energy as a radio wave, an antenna is connected to the tuned circuit as shown in the figure.

To build our oscillator/transmitter the layout in figure 73 should be followed and all connections carefully checked. Then a temporary an-

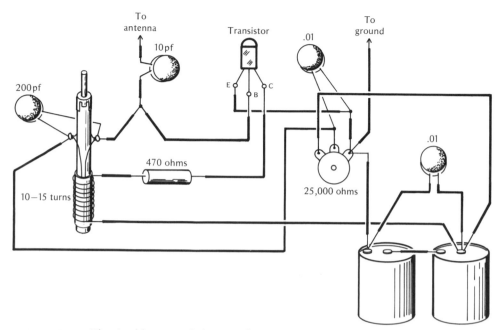

FIGURE 73 *Physical layout of the simple oscillator.*

tenna can be made of 2–3 feet of insulated wire. For initial testing take a small radio and tune it to a spot on the dial where there are no stations. Apply power to the oscillator and turn the variable resistor halfway between maximum and minimum setting.

Slowly tune the variable coil until you hear the signal on the receiver: a hissing noise which sounds like a station with no talking—which is exactly what it is. Now readjust the variable resistor to the most stable operating point.

The circuit can be used to transmit telegraphic signals by connecting a telegraph key in series with the battery (figure 74). Try this—it takes only a moment—and you have a true radiotelegraphic transmitter.

You can experiment with a longer antenna: you may have to readjust the variable resistor. The longer the antenna, the more radio energy is absorbed by it and the less is available for feedback. A 20–50-

FIGURE 74

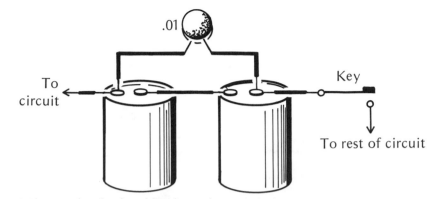

Where to hook a key for telegraph transmissions.

foot antenna allows an operating range, with a sensitive receiver, of a few hundred feet, or about the limit of what can be obtained with such low-power equipment (although you can certainly experiment with longer or higher antennas to attempt to increase the range).

To send voice signals over the transmitter, connect a carbon microphone to the emitter lead of the transistor (figure 75). The operating current of the transistor will vary directly in step with incoming sound waves, and will, in turn, vary the strength or amplitude of the radio waves being produced so that speech will be transmitted. The variable resistor should be readjusted for best overall clarity.

To further help increase the distance that can be covered, the

FIGURE 75

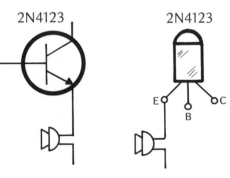

Where to add a carbon microphone to the oscillator for voice transmissions.

FIGURE 76

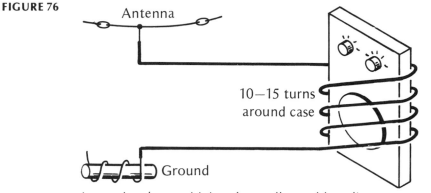

Increasing the sensitivity of a small portable radio.

receiver can be modified by connecting it to an outside antenna. Connect the lead-in wire to 10–15 turns of wire wound around the case of the receiver (figure 76). The other end of the coil should be connected to ground. This will increase the number of commercial stations received; a new quiet spot on the dial may have to be found.

Careful tuning and good antennas should allow a range of up to 1000 feet. As an added convenience, a switch can be wired (figure 77) to allow the use of the same antenna for transmitting and receiving.

The main factor limiting the range of our transmitter is the amount

FIGURE 77

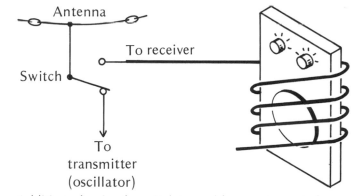

Addition of a simple switch to enable an antenna to be used for transmitting or receiving.

of radio energy it can produce. To operate over greater distances requires much more power, which is obtained by amplifiers. Commercial broadcasting stations use huge amplifiers that consume thousands of watts of power and antennas that are towers hundreds of feet high. In New York City, for example, the commercial stations have antennas on top of skyscrapers and pump 50,000 watts of power into them. Broadcasts can be heard 30–50 miles away with small, simple receivers, and hundreds of miles away with sophisticated receivers using rooftop antennas.

CHAPTER 10
AMATEUR RADIO

If at this point the wireless communications bug has really bitten, there is only one path to follow and that is to become an amateur radio operator, to build transmitting and receiving equipment by which you can talk to people all over the world. You can meet people who share your interests and can take part in worldwide contests, emergency or disaster communications, even technological breakthroughs. Most important, you will experience the thrill of talking to someone hundreds or thousands of miles away by means of equipment you have built with your own hands. The memory of this accomplishment will never leave you.

To become a radio amateur you will need a license from the Federal Communications Commission. The FCC, founded in 1934 to eliminate the confusion that would arise if everyone tried to transmit on the same frequency, assigns all frequencies in use within the United States, including those that amateur radio operators may use.

It is not difficult to obtain an FCC license: people from age 7 to well over 80 have done it. The applicant must learn and demonstrate certain skills, and be familiar with the rules and regulations pertaining to amateur radio, as well as some elementary electronics theory. One skill that must be mastered is the sending and receiving of international Morse code. If you have built the telegraph set in this book and spent much time with it, you probably already have this skill. If not, practice (with a friend, if possible) until you can send and receive at least five five-letter words per minute. This, plus the knowledge of some basic rules, regulations, and theory, will enable you to obtain a novice license. This permits you to build simple equipment and do enough communicating to increase your code speed and theory to the point where you can get a general license. Then the world will be literally at your fingertips.

The best way to study for an amateur radio license is to obtain a workbook/tape cassette course from the Amateur Radio Relay League (ARRL), a nonprofit organization. Since 1914, the ARRL has been the traditional spokesman for the amateur and is dedicated to promoting ham radio. It is owned, directed, supported, and run by amateurs—most of them elected by members.

The course is inexpensive and covers, in a logical way, the steps necessary to obtain a license. It shows construction details for simple, inexpensive but higher-power transmitters and more sophisticated receivers than those discussed in previous chapters. The equipment is intended for use on amateur radio frequencies, and some of it is powerful enough for worldwide communications.

Another way to become an amateur radio operator is to join a local radio club. Many schools, churches, and youth groups have such clubs and give courses on amateur radio. You will meet people who have the same interests and goals as you and the procedure of getting your license will be that much more fun. Another advantage of a radio club is that you can trade for or purchase equipment you need from other experimenters. Many amateur radio operators are experimenters and have large collections of components and unused equipment that they are only too happy to trade, give away, or sell. Flea markets run by these clubs attract thousands of interested experimenters.

FIGURE 78

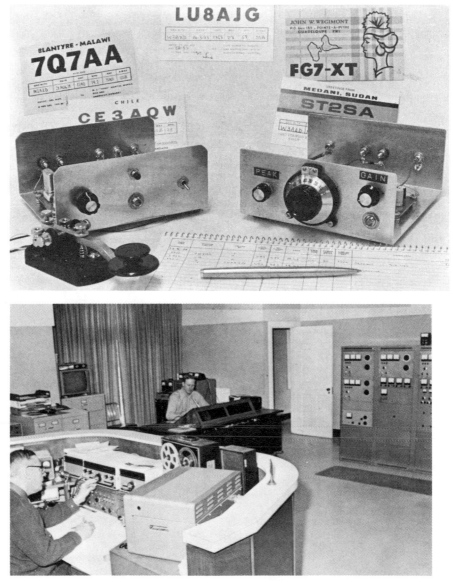

Simple and complex amateur radio stations. Either is capable of achieving worldwide communications.

When you have completed your studying, taken your exam, and passed, you will soon receive your license. On the license will be a call sign designating your station and identifying you as a member of the fraternity of amateur radio operators. Then, after building a transmitter and receiver and stringing an antenna, you will be ready to operate as an amateur.

As the final switch is thrown and you press down the key or speak into the microphone, your signals may flash over thousands of miles to a similarly equipped station, perhaps in another country. Switching to "receive," you will hear the never-to-be-forgotten voice of the distant station calling you back. This is what amateur radio is all about.

As you gain experience talking to many other amateurs your skill and knowledge will increase, and the thrill of being able to talk around the world will never wear out.

for further reading

Beginner's Level

Electricity and Electronics by Howard H. Gerrish (South Holland, Ill.: Goodheart-Willcox)

Understanding Radio (3rd ed.) by H. M. Watson, H. E. Welch and G. S. Eby (New York: McGraw-Hill)

Intermediate Level

Transistor Electronics by Howard H. Gerrish (South Holland, Ill.: Goodheart-Willcox)

General Level

Technical Electricity and Electronics by Peter Buban and Marshall L. Schmitt (New York: McGraw-Hill)

Publications of the American Radio Relay League, Newington, CT 06111 (Write for publications list.)

Advanced Level

Electronic Communication (3rd ed.) by Robert L. Shrader (New York: McGraw-Hill)

Periodicals

Radio Electronics (Gernsback Publications, 200 Park Avenue South, New York, NY 10017)

Popular Electronics (Ziff-Davis, One Park Avenue, New York, NY 10016)

Ham Radio Horizons (Greenvale, NH 03048)

CQ (Cowan Publishing Corp., 14 VanDerventer Avenue, Port Washington, NY 11050)

QST (ARRL, Newington, CT 06111)

(*Ham Radio Horizons, QST,* and *CQ* carry articles of interest to the beginner.)

index